Ruth Hobart

MOWGLI
learns to swim

Loosely based on
Rudyard Kipling's **The Jungle Book**

商務印書館

CONTENTS

Originally published by Black Cat Publishing under the title: *Mowgli learns to swim*
© 2000 Black Cat Publishing
An imprint of Cideb Editrice, Genoa, Canterbury

The copyright of this Chinese edition is owned by
The Commercial Press (H.K.) Ltd.

Name of Book: Mowgli learns to swim
Author: Ruth Hobart
Editors: Monika Marszewska, Claudia Fiocco
Design and Art Direction: Nadia Maestri
Illustrations: Alfredo Belli
Layout: Sara Blasigh

系 列 名：Quality English Learning for Kids · I
書　　名：Mowgli learns to swim
責任編輯：傅　伊
出　　版：商務印書館（香港）有限公司
　　　　　香港筲箕灣耀興道 3 號東滙廣場 8 樓
　　　　　http://www.commercialpress.com.hk
印　　刷：美雅印刷製本有限公司
　　　　　九龍觀塘榮業街 6 號海濱工業大廈 4 樓 A
版　　次：2004 年 9 月第 1 版第 1 次印刷
　　　　　© 2004 商務印書館（香港）有限公司
　　　　　ISBN 962 07 1721 X
　　　　　Printed in Hong Kong

出版説明

　　學英語當然要學優質的，有品質才能讓人有信心。我們一直積極提倡學習優質英語的理念，並且為學習者提供過多元化的優質英語材料，像《Black Cat 優質英語階梯閱讀》就十分成功，至今已出版近 60 本。鑑於良好的英語能力最好從小培養，我們於是出版這一套適合五至八歲兒童的優質英語閱讀讀本 "Quality English Learning for Kids·I"。

　　培養兒童對於英語的興趣，須從趣味和簡易兩方面入手。圖文並茂，聲文結合這兩大特點對學習英語甚有幫助。"Quality English Learning for Kids·I" 承續本館出版優質英語書的理念，全書彩圖精美，附 CD 朗讀內容及聆聽練習，形式多元化，有出版故事讀本（story books）、圖畫讀本（picture readers）、戲劇讀本（drama readers）及互動讀物（interactive readers）四大類，提供不同學習功能。故事讀本和圖畫讀本可供兒童看圖講故事；戲劇讀本完全用對白編寫，培養脫口而出講英語的習慣，適合家庭裏作簡單的角色扮演，或者小學生在課堂上作簡單的演出。

　　針對兒童學習英語的需要，本系列提示家長為兒童設定學習目標，並且說明如何達標，另備生詞表和語法知識點，讓兒童在家長協助下掌握生詞用法，認識簡單的句子結構和了解語法要點。

　　"Quality English Learning for Kids·I" 吸引兒童對閱讀產生興趣，逐步引導他們參與愉快的閱讀旅程。在這個旅程中，家長是重要的導航者，透過對兒童的悉心鼓勵，循循善誘，進一步加強親子關係。

商務印書館
編輯部

使用説明

1 如何使用本書？

本書為故事讀本（story reader），適合課堂使用或親子共讀。

每頁均圖文並茂。正文包括對話和敍述文字，讓小孩子熟悉第一、二、三身句法。老師或家長可與小孩子展開簡單對話，討論朋友、年齡、姓名、喜歡的活動等話題。老師或家長還可讓小孩子使用祈使句發出指示。

除正文外，還設有問答、選擇填空、判斷正誤等練習題。老師或家長可指出常見食品和動物，讓小孩子説出它們的名稱和顏色。

本書配有 CD，小孩子可邊聽邊讀，提高英語聽説能力。

2 本書的學習目標是甚麼？

老師或家長可為孩子定出以下學習目標。

使用本書後，孩子學會：

(a) 説出動物的名稱（say the names of animals）；

(b) 説出某物放在甚麼地方（say where things are）。

3 本書有哪些重點生詞和語法學習項？

(a) 重點生詞：本書的重點生詞包括三大類，即動物（animals）、食品（food）、身體部位（parts of the body）。另附圖典（pictionary）和英漢對照生詞表（glossary），幫助理解和記憶生詞。

(b) 語法學習項：

第一身句法（the first person）（例如頁 9，"I'm hungry!"）

第二身句法（the second person）（例如頁 15，"You're a big boy, Mowgli."）

第三身句法（the third person）（例如頁 8，"They sit on a rock to rest."）

can 用於表示某人具有某種能力，置於動詞原形前面，即 "can＋動詞原形"（can + verb）（例如頁 15，"Boys of 8 can swim."）；表示某人不具備某種能力時，可用 "can't＋動詞原形"（can't + verb）結構。（例如頁 7，"I can't swim!"）

簡單現在時（present simple）動詞為原形，但當主語為第三身單數名詞或代詞時，動詞為 has、is 等特殊形式或在詞尾加 s。（例如頁 4，"A big tiger growls!" 和頁 6，"Bagheera and Baloo jump into the river."）

祈使句（imperatives）主要由 "動詞＋賓語"（verb + object）組成。（例如頁 6，"Jump in!" 和頁 9，"Look at Shere Khan!"）

介詞（prepostions）用於表示地點，後面接名詞。（例如頁 3，"...near the river." 和頁 5，"The tiger is behind the tree!"）

PART ONE

2 Mowgli, Bagheera the panther and Baloo the bear are having a picnic near the river.

A big tiger growls!
GRRRRRRRR!

The tiger is behind the tree!

Bagheera and Baloo jump into the river.
The water is cold!
'Jump in! Jump into the river Mowgli!' they shout.

'No! No! I can't swim! I can't swim!' says Mowgli.
'Jump onto my back!' shouts Bagheera.
Mowgli jumps onto Bagheera's back.
Bagheera, Baloo and Mowgli swim across the river.

Bagheera and Baloo are tired!
They sit on a rock to rest.
'Who is he?' asks Baloo.
'His name is Shere Khan the tiger,' answers
Bagheera.

QUALITY ENGLISH CLUB

The Commercial Press (Hong Kong) Ltd.
8/F, Eastern Central Plaza,
3 Yiu Hing Road, Shau Kei Wan,
Hong Kong

THE COMMERCIAL PRESS (H.K.) LTD.

QUALITY ENGLISH CLUB
Membership Application Form

QUALITY ENGLISH CLUB is for those who love English reading and seek for better English to share and learn with fun together.

Benefits offered: - *Membership Card*   - *English learning activities*

   - *English learning e-forum*   - *Surprise gift and more...*

Simply fill out the application form below and fax it back to 2565 1113 or send it back to the address at the back.

Join Now! It's FREE exclusively for readers who have purchased books on Quality English Learning published by the Commercial Press!

(Please fill out the form with **BLOCK LETTERS**.)

The title of book(s) /book set(s) that you have purchased: _____

English Name: _____ (Surname) _____ (Given Name)

Chinese Name: _____

Address:

Tel: _____   Fax: _____

Email: _____

(Login password for e-forum will be sent to this email address.)

Sex: ❏ Male   ❏ Female

Education Background: ❏ Kindergarten   ❏ Primary 1-3   ❏ Primary 4-6
   ❏ Junior Secondary Education (F1-3)   ❏ Senior Secondary Education (F4-5)
   ❏ Matriculation   ❏ College   ❏ University or above

Age: ❏ 3 - 6   ❏ 6 - 9   ❏ 10 - 12   ❏ 13 - 15   ❏ 16 - 18
   ❏ 19 - 24   ❏ 25 - 34   ❏ 35 - 44   ❏ 45 - 54   ❏ 55 or above

Occupation: ❏ Student   ❏ Teacher   ❏ White Collar   ❏ Blue Collar
   ❏ Professional   ❏ Manager   ❏ Business Owner   ❏ Housewife
   ❏ Others (please specify: _____)

As a member, what would you like **QUALITY ENGLISH CLUB** to offer:

❏ Member gathering/ party   ❏ English class with native teacher   ❏ English competition
❏ Newsletter   ❏ Online sharing   ❏ Book fair
❏ Book discount   ❏ Others (please specify: _____)

Other suggestions to **QUALITY ENGLISH CLUB**: _____

_____

Please sign here: _____   (Date: _____)

Visit us at Quality English Learning Online http://publish.commercialpress.com.hk/qel

'Look at Shere Khan!' says Bagheera.
'He *must not* eat the picnic!'
'Yum! Yum! I'm hungry!' says the tiger.
'Delicious ham sandwiches! Fantastic sausage rolls and orange juice... Yum! Yum! Yum!'

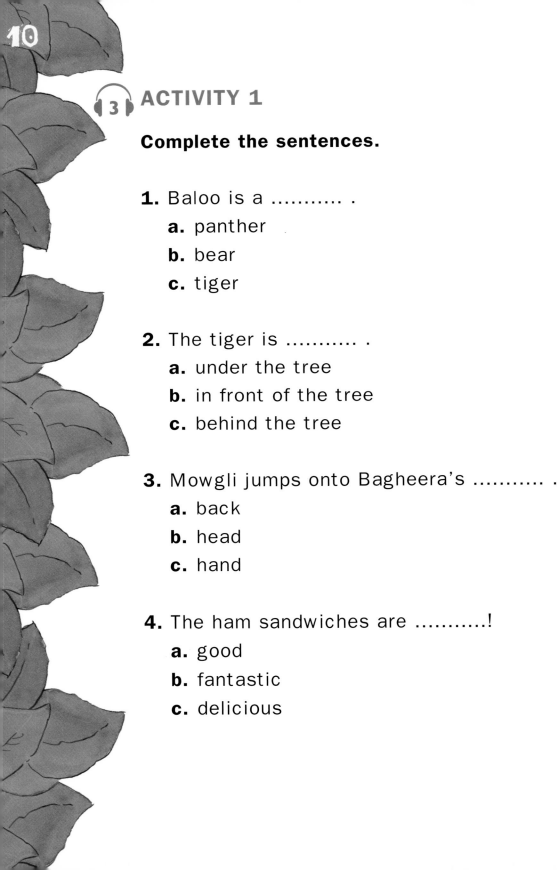

3 **ACTIVITY 1**

Complete the sentences.

1. Baloo is a ....
   a. panther
   b. bear
   c. tiger

2. The tiger is ....
   a. under the tree
   b. in front of the tree
   c. behind the tree

3. Mowgli jumps onto Bagheera's ....
   a. back
   b. head
   c. hand

4. The ham sandwiches are ....!
   a. good
   b. fantastic
   c. delicious

ACTIVITY 2

Answer YES or NO.

1. Are Mowgli and his friends having a picnic?

....

2. Is the tiger Mowgli's friend?

....

3. Can Baloo and Bagheera swim?

....

4. Is the tiger's name Shere Khan?

....

5. Is the tiger hungry?

....

PART TWO

Mowgli, Bagheera and Baloo are very sad.
The picnic is in the tiger's tummy!

'What's the matter?' asks Hathi, the elephant.
'Look at the greedy tiger!' says Mowgli.
'My delicious picnic is in the tiger's tummy!'

'Go back across the river!' says Hathi.
'Go back to the tiger and say: Naughty, naughty
tiger! You *must not* eat the picnics any more!'
'No, I can't go back! I can't go back!' says Mowgli.
'I can't swim!!!'

'How old are you Mowgli?' asks Jo the monkey.
'I'm 8, I think,' says Mowgli. 'Yes, I am. I'm 8!'
'You are a big boy, Mowgli. Boys of 8 can swim,'
says Mandy the mouse.
'I can't swim!' says Mowgli.

'I can teach you to swim!' says Hathi.
'I can teach you to swim!' says Jo the monkey.
'I can teach you to swim!' says Bagheera.

'I can teach you to swim!' says Baloo.
'I can teach you to swim!' says Mandy the mouse.
'Thank you! Thank you!' says Mowgli.

Baloo, Bagheera, Jo and Mandy jump into the river!
'Are you ready Mowgli?' they shout.
'Are you ready to jump into the river?'
One! Two! Three! Ready, steady GO!!!
Splash! Mowgli jumps into the river!

ACTIVITY 3

Complete the sentences.

1. Mowgli, Bagheera and Baloo are
very ....
  a. happy
  b. sad
  c. thirsty

2. Mowgli is ....
  a. seven
  b. six
  c. eight

3. Jo is ....
  a. a mouse
  b. an elephant
  c. a monkey

4. Mowgli .... into the river.
  a. jumps
  b. runs
  c. walks

(7) ACTIVITY 4

Answer True or False.

1. The picnic is in the tiger's tummy.

....

2. The tiger is good.

....

3. Mowgli can swim.

....

4. The animals can teach Mowgli to swim.

....

5. The animals jump into the ocean.

....

PART THREE

'Good! Good boy, Mowgli!' say the animals.
'Move your legs up and down, up and down!' says Jo.
'Move your arms in and out, in and out!' says Mandy.
'Close your mouth and open your eyes!' says Baloo.
'Move your arms and legs in and out and up and down!' says Bagheera.

'Help!' shouts Mowgli. 'Help! I can't swim!'
'Climb onto my trunk,' Hathi tells Mowgli.
Mowgli is on Hathi's trunk!
'Ha! Ha! Ha!' laugh Baloo, Bagheera, Mandy and Jo!
'How are you Mowgli?' they ask.
'I'm fine!' shouts Mowgli. 'I think I can swim,
but not very well!'

'Jump into the water. Jump into the water and swim!' says Hathi.
'Yes! Look!' says Mowgli. 'I can jump into the water and I can swim!'
'Let's sing a song!' shout Mandy and Jo.

Baloo sings with the other animals.
'You can swim, you can swim!
Yes you can, yes you can!
Jump into the river, jump into the river!
Splash! Splash! Splash!
Splash! Splash! Splash!'

'Let's swim back to the naughty tiger!' shouts
Mowgli.
'The naughty tiger *must not* eat the picnics any more!'
'Yes!' shouts Baloo.
'Let's go back!' says Bagheera.

'Hooray! Hooray!' shout the monkeys, zebras,
giraffes, lions and elephants.
They all clap and clap.
'Very good Mowgli. You can swim now!!!'

'Where is the tiger?' asks Mowgli.
'The naughty tiger is in the forest,' say the animals.
'Let's have a party!' says a lion.
'We have a lovely picnic for you!' says a giraffe.

ACTIVITY 5

🎧 9 Write the correct word.

in   open   up
out   down

1. Mowgli's arms go .... and ....

2. Mowgli's legs go .... and ....

3. Mowgli's eyes are ....

🎧 10 ACTIVITY 6

Mowgli can swim now.
Draw a picture of Mowgli in the river.

🎧 ACTIVITY 7

**Find the words
in the word square.**

mouse

Mowgli

Baloo

monkey

B	A	G	H	E	E	R	A	Z	W
A	C	I	Q	T	E	M	B	O	P
L	U	R	L	M	O	W	G	L	I
O	M	A	L	I	F	H	J	K	N
O	O	F	Y	W	O	U	S	M	Q
E	U	F	G	Z	I	N	L	O	M
P	S	E	L	E	P	H	A	N	T
R	E	T	V	B	X	S	B	K	U
T	I	G	E	R	T	L	O	E	B
J	B	V	D	A	C	M	S	Y	N

lion

Bagheera

tiger

elephant

zebra

giraffe

MOWGLI
learns to swim

KEY TO THE EXERCISES

Page 10
ACTIVITY 1
1. bear
2. behind the tree
3. back
4. delicious

Page 11
ACTIVITY 2
1. Yes
2. No
3. Yes
4. Yes
5. Yes

Page 19
ACTIVITY 3
1. sad
2. eight
3. a monkey
4. jumps

Page 20
ACTIVITY 4
1. True
2. False
3. False
4. True
5. False

Page 28
ACTIVITY 5
1. in / out
2. up / down
3. open

Page 29
ACTIVITY 7

B	A	G	H	E	E	R	A	Z	W
A	C	I	Q	T	E	M	B	O	P
L	U	R	L	M	O	W	G	L	I
O	O	A	L	I	F	H	J	K	N
O	M	F	Y	W	O	U	S	M	Q
E	O	F	G	Z	I	N	L	O	M
P	U	E	L	E	P	H	A	N	T
R	S	T	V	B	X	S	B	K	U
T	I	G	E	R	T	L	O	E	B
J	B	V	D	A	C	M	S	Y	N

PICTIONARY

animals

arms

back

bear

behind

boy

clap

down

elephant

eyes

forest

giraffe

growls/grrr

ham sandwiches

jump

laugh

legs

lion

monkey

mouse

mouth

one

onto   on

orange juice

panther

picnic

river

rock

sad

sausage rolls

sing

sit

Splash!

swim

three

tiger

tired

tree

trunk

tummy

two

up

yum!

zebra

GLOSSARY

animals 動物

arms 胳臂

back（人體）背部

bear 熊

behind 在…的後面

clap 拍手

delicious 美味的

elephant 象

eyes 眼睛

fantastic 極好的

forest 森林

giraffe 長頸鹿

greedy 貪婪的

growl 狂叫

ham sandwiches 火腿三明治

hungry 饑餓的

jump 跳躍

laugh 笑

legs 腿

lion 獅子

monkey 猴子

mouse 老鼠

mouth 嘴

move 移動，擺動

naughty 調皮的

on 在…的上面

onto 移到…上

orange juice 橘子汁

panther 豹子

picnic 野餐

rest 休息

river 河流

rock 岩石

sad 傷心的

sausage rolls 香腸卷

shout 叫喊

splash 水濺潑聲

swim 游泳

teach 教學

thirsty 口渴的

tiger 老虎

tired 疲勞的

tummy 肚子

up 向上，向較高處

zebra 斑馬

USEFUL EXPRESSIONS

Are you ready? 你準備好了麼？

Go back... 回去…

Good boy! 好孩子！

Help! 救命！

His name is... 他的名字是…

Hooray! 好哇！

How are you? 你好麼？你身體怎麼樣？

How old are you? 你多大年紀了？

I can swim. 我會游泳。

I can't swim. 我不會游泳。

I can teach you to swim. 我可以教你游泳。

I'm fine! 我很好。

I'm hungry! 我餓了。

Let's have a party! 我們舉行一次聚會吧！

Look at Shere Khan. 瞧 Shere Khan。

Not very well. 不是很好。

Ready, steady, go!!! 預備，站穩——跳！！！

Thank you! 謝謝你！

Very good. 很好。

What's the matter? 發生甚麼事了？

Where is...? …在哪裏？

Who is he? 他是誰？

Yes. 是。

You must not... 你絕不能…